Edward Oram Shakespeare

A Criticism of Doctor Formad's Printed Statements and Conclusions Concerning the Aetiology of Tuberculosis

Edward Oram Shakespeare

A Criticism of Doctor Formad's Printed Statements and Conclusions Concerning the Aetiology of Tuberculosis

ISBN/EAN: 9783337865849

Printed in Europe, USA, Canada, Australia, Japan

Cover: Foto ©berggeist007 / pixelio.de

More available books at **www.hansebooks.com**

Reprinted from the New York Medical Journal for Aug. 9 and 16, 1884.

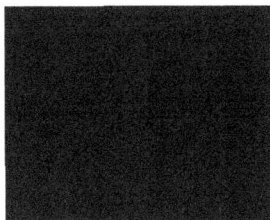

A CRITICISM OF

DR. FORMAD'S PRINTED STATEMENTS AND CONCLUSIONS CONCERNING THE ÆTIOLOGY OF TUBERCULOSIS.

By EDWARD O. SHAKESPEARE, A. M., M. D.,

PHILADELPHIA,

OPHTHALMIC SURGEON AND PATHOLOGIST TO THE PHILADELPHIA (BLOCKLEY) HOSPITAL, ETC.

THE "New York Medical Journal," in its issue of June 14th of this year (p. 675), contains a report, substantially correct, of the remarks of Dr. Shakespeare during the discussion before the Philadelphia County Medical Society of Dr. Formad's second paper on "The Ætiology of Tuberculosis." In that discussion the remarks of Dr. Shakespeare were limited to three points: the statements made by Dr. Formad of Koch's animus and present views, and a consideration of Dr. Formad's theory concerning the condition of the lymph-spaces in the scrofulous.

In the issue of the same journal of June 28th appears (p. 723) a quite characteristic letter from Dr. Formad, written from the Pathological Laboratory of the University of Pennsylvania. In that letter are certain disclaimers and denials of responsibility for misrepresentations of Koch's present views and animus, with which he was charged dur-

ing the above-mentioned discussion before the Philadelphia
Medical Society.

It is to be remarked that it is a little odd that these disclaimers and denials have been addressed to the readers of
the "Journal" rather than to the members of the society before whom his address was delivered. Perhaps, however,
those who heard Dr. Formad as he delivered his address,
seeing him refer only occasionally to the notes at his side,
and also heard at the close of the debate a most positive
reiteration of the statements which had been criticised by
Dr. Shakespeare, can readily appreciate the reasons which
may have caused the author of the letter to address a denial
to the readers of the "Journal" rather than to themselves.
If, after mature reflection, aided no doubt by the recent appearance of the second volume of the "Mittheilungen aus
dem kaiserlichen Gesundheitsamte," in which Koch has
declared his present views concerning the tubercle bacillus,
Dr. Formad has found it necessary to retract what he said
before his audience in Philadelphia in misrepresentation of
those views, it would have been manly, honorable, and
praiseworthy for him to do so in a straightforward manner.
But he has preferred to deny his words, and to address
this denial to those who did not hear them. Probably
those few of the readers of the "Journal" who did hear
the author's words, and have also read the printed article,
will not only be struck with the vast difference between
some of the oral and the printed statements, but also will
be surprised at the declaration that the quotation given in
his letter contains *all* that he has ever spoken or published
upon that point. The remarks which Dr. Shakespeare
made before the Philadelphia society during the debate
were in criticism of statements *orally addressed to the ears*
of that body, and not in criticism of what had been for the
most part rewritten and prepared for publication many

weeks after the address was delivered. But, unfortunately for the author of the letter, there is sufficient basis, even in the greatly modified and tempered language which he quotes as his own, upon which to ground a charge of misrepresentation.

The charge which the author of the letter denies and to which, in his own language, he pleads "*not guilty*," is substantially reported by the recorder as follows:

"2. The author had further announced that Koch had so far modified his views that he now admitted that neither the form, size, and aspect of the tubercle bacillus, nor its want of individual motion, nor its peculiar behavior toward staining-fluids, distinguished it from many other bacilli." ("New York Medical Journal," June 14, 1884, p. 675.)

The language quoted by our author as his genuine and only utterance, and which he thinks to justify his denial, comprises the following sentences: "Koch seems now to lay more stress upon this low-power appearance (the S shape of the culture colonies), and upon the pathogenic properties of the bacillus tuberculosis, as a distinguishing feature from other bacilli, than upon the color test. During the conversation he admitted that some other bacilli may also yield the same micro-chemical reaction as the tubercle bacilli, but insisted that the latter can not be stained brown."

It may not be necessary to reproduce here the actual words in which Koch has expressed his present views, since they have probably by this time been read by every one interested in them. It may, however, be profitable, for more purposes than the present, to render the following into English, as an offset to the representations of the author of the letter:

Speaking of the peculiar action of Ehrlich's special staining method upon the tubercle bacilli and their surroundings, Koch says: "But in the differentiation of the tubercle ba-

cillus this peculiarity (staining the tubercle bacilli a color
in contrast to that of their surroundings) renders still further assistance, for not alone do the connective-tissue con-
stituents assume the contrast color, but all other bacteria at
present known to me, except the lepra bacillus, later to be
mentioned, likewise stain by Ehrlich's coloring method in
contrast to the tubercle bacilli. . . . Recently I have tested
Ehrlich's method of staining upon numerous substances
containing bacteria, such as putrid meat-infusion, decom-
posing urine, blood, and milk, foul vegetable infusions,
swamp-slime, etc., but have never found bacteria which gave
the same color-reaction as the tubercle bacilli. I must
therefore regard all assertions concerning the occurrence of
bacteria which may have been found in sputum, in putrid
fluids, intestinal contents of healthy men, swamp-slime, and
stained like the tubercle bacilli, as erroneous and dependent
upon a faulty application of the method of staining." (" Mittheilungen aus dem kaiserlichen Gesundheitsamte," zweiter
Band, Berlin, 1883, p. 12.)

" Besides the tubercle bacilli, there is at present only one
single species of bacterium known which colors in a manner similar to the tubercle bacilli. This is, as I mentioned
in my first communication, the lepra bacillus. This fact is
the more noteworthy since not only are the parasites of tuberculosis and lepra in manifold respects similar and apparently nearly related, but it is also known that both diseases,
even anatomically as well as ætiologically, stand very close
together. Of course, the staining qualities of both are not
perfectly identical. For, while the lepra bacilli can be col-
ored by the same staining fluid as can the tubercle bacilli,
yet the reverse is not the case. The former, as is known,
imbibe Weigert's nucleus-coloring material, as Neisser first
pointed out, while the latter do not. Notwithstanding the
great similarity of these two species of bacilli, still one is

able to differentially diagnose each of these two forms of bacilli by means of their different behavior toward Weigert's nucleus-coloring fluid." (*L. c.*, p. 13.)

It is by no means our intention to enter upon an exhaustive criticism of those papers. Neither does time or inclination serve for such a task, for, judging from the abundance of material presented, it might prove almost interminable. Yet the position which their author occupies in this country as the demonstrator of pathology and lecturer on experimental pathology in one of the greatest of our medical schools would seem to entitle him to speak with authority upon the subject of which he writes, and would also seem to enable him to exert no little influence in the development of public opinion concerning those matters. Moreover, he claims to be a leader among the few who, from the standpoint of personal observation and experiment, have denounced Koch's claims for the agency of the tubercle bacillus in the ætiology of tuberculosis, as the following expression appears to indicate : "The only men who attempted to repeat Koch's experiments, besides the work done in the Pathological Laboratory of the University of Pennsylvania, were Spina and Watson Cheyne." ("Philadelphia Medical Times," Feb. 9, 1884, p. 339.) Furthermore, in the stand which he has taken with regard to the question of the ætiology of tuberculosis the author is undoubtedly upon the side of prevalent professional opinion in America, and perhaps also of widespread popular belief.

All these considerations would appear to justify a careful and extended review of our author's knowledge and opinions concerning the ætiology of so fatal and widespread a disease as tuberculosis. Besides, the incalculable importance to the whole human race, for this and all future time, of Koch's discoveries, if they be confirmed ; the skill, reliability, and experience of Koch in the investigation of minute organisms which may be the germs of disease; and the abundant, almost overwhelming array of strong, exact, and convincing proof, both original and confirmatory, which has been brought to the support of this new theory—make it the bounden duty of every physician, who would properly regard his obligations, not only to searchingly scrutinize the proofs upon which this theory rests, but also at the same time to most rigidly examine the ground upon which its antagonists stand.

There is, therefore, no occasion to apologize for presenting either a partial or a complete examination of Dr. Formad's opinions concerning tuberculosis.

In the course of his remarks in discussion of the author's last paper, as appears by the record published in the "Journal," the writer took occasion to state that there were very many points assumed as demonstrated, and positive statements advanced in the elaboration of this paper, which he believed to be without sufficient foundation. The letter of Dr. Formad has already been spoken of as characteristic. It is in some degree also characteristic of his already published papers upon tuberculosis, as will perhaps shortly appear.. In the following incomplete review, by quoting the author against himself in refutation of some of his important assertions and conclusions, and by quoting the words of other authors in refutation of assertions and conclusions for which he has made them responsible, the endeavor will be made to show that our author has not only misrepresented other authors than Koch, but that he has repeatedly, even in the most important matters, misrepresented and flatly contradicted himself. Furthermore, it will be shown, by evidence which the author can not successfully repudiate, that he has misrepresented the extent and character of observations reported. Fully appreciating the gravity of such charges against the publications of one who would authoritatively treat of the ætiology of such a disease as tuberculosis, we now proceed, as far as space and time will permit, to substantiate them.

Toward the end of his second paper on Tuberculosis, Dr. Formad submits " a brief analysis and summary of experiments made and evidence offered in relation to the question of the parasitic origin and specific nature of tuberculosis."

Apropos of this, he continues as follows : " For the establishment of a theory in regard to a parasitic origin of a disease by means of experiments upon animals, etc., the following propositions must be affirmatively decided :

" *1. The disease produced experimentally in animals by means of inoculation with products of human disease must be proved to be identical with the disease occurring spontaneously in man.*" (" Phila. Med. Times," April 5, 1884, p. 488, right column, par. 3.)

The author thus concludes the consideration of this proposition : " The proof, then, upon this point—the supreme one for the settlement of the question of the nature of tuberculosis—is yet to be furnished." (*L. c.*, p. 490, par. 1.)

What are the grounds upon which this conclusion is based, and how is it arrived at ?

In partial conflict with the foregoing conclusion are the following words of the author found in the sentence immediately preceding : " Judging from my own experiments, there is to my mind no doubt that some forms of artificially · induced tuberculosis in animals acquire gradually characters which make them identical with spontaneous tuberculosis in man or beast." (*L. c.*, p. 489, right column, par. 5.) In complete conflict with this conclusion are the following statements of our author : " In my previous studies, judging from the literature alone, I was fully impressed with the idea that tuberculosis had a specific exciting cause, and that it could be induced by inoculation with tuberculous materials. Moreover, having made numerous inoculations with tuberculous matter, I convinced myself of this fact. . . . But at the same time, after repeating, under various modifications, the well-known control experiments, I found that, beyond doubt, even true tuberculosis could be produced by substances other than tuberculous." (*L. c.*, p. 493, right column, pars. 4 and 5.)   " There is no doubt that Koch's tubercle bacillus, when isolated and cultivated for many generations and then inoculated into certain animals, is capable of inducing tuberculosis, or a nodular eruption not distinguishable from it, more readily than other irritants, so far as tried." (*L. c.*, p. 491, right column, par. 4.)   " Although the tubercle bacillus is more liable to excite tuberculosis in an already inflamed and ill-nourished soil than all other simple irritants so far tested, it (the bacillus) might be readily substituted by other irritants." (*L. c.*, p. 493, left column, par. 2.)

We fail to rightly understand the meaning of language and the common methods of ratiocination and sequence of thought if the man who uttered these expressions did not believe in the undoubted identity of human tuberculosis and that produced experimentally in animals. In further conflict with this conclusion is the following statement of the author : " Spontaneous animal tuberculosis is unquestionably identical with human tuberculosis. There are a few morphological specializations which I mentioned in a former chapter—e. g., in tuberculosis of birds and in bovine tuberculosis, or pearl disease ; but the essential peculiar histological features are the same in all. Tubercle bacilli appear also to be present in nearly all cases of spontaneous animal tuberculosis. I detected bacilli in a tuberculous bronchial lymph-gland from a phthisical tiger, which I had kept in alcohol for eight years, and in one from a monkey of more recent date ; and several times I found bacilli in spontaneous bovine, chicken, rabbit, and guinea-pig tuberculosis. I also studied tuberculosis in the bear, lion, leopard, and in a large variety of apes (dead of phthisical consumption from the Zoölogical Garden), with results identical with those obtained from studies in man. But this was long before the outbreak of the *bacilliary campaign*, and, consequently, Koch's parasite was not looked for in these latter cases." (*L. c.*, March 22, p. 453, right column, par. 5.) Possibly the rabbits which the author has so frequently chased may be responsible for this extraordinary doubling.

It may, perhaps, not be necessary to go further into an examination of the manner in which the conclusion as to the first proposition has been reached. But it may be at least interesting, and possibly also instructive, to the plain, straightforward, common-sense student of medicine to learn something more of the mysterious ways by which an uncommon mind grasps conclusions.

Our author views his first proposition from three different standpoints, in the order here mentioned, viz., inoculation, inhalation, feeding. Let us for convenience reverse this order and commence with the last, viz., *feeding*.

We quote one of our author's arguments in support of his already-mentioned conclusion : "The following deserves a passing mention. According to Orth (Virchow's 'Arch.,' vol. lxxvi) and Bollinger ('Arch. f. exper. Path.,' vol. i), there is some doubt as to the identity of human and animal tuberculosis."

Let us see whether or not these authors are responsible for that statement.

Professor J. Orth ("Experimentelle Untersuchungen über Fütterungstuberculose," Virchow's "Archiv," vol. lxxvi, p. 234), says : "I hesitate not for this affection [he is speaking of the disease produced in animals by feeding with tuberculous matter] to claim the name of tuberculosis. These irregularly eroded ulcers which proceed from caseous degeneration of minute nodules, these cheesy nodes formed by the confluence of cheesy nodules, these minute, isolated nodules themselves, with their reticulum and their many nucleated giant-cells—what else should they be other than tubercular ulcers, tubercular aggregations, isolated young tubercles, even if they are compared with human tuberculosis?"

And again, speaking of the constitution of nodules (p. 235), Orth says : ". . . all the nodules in my rabbits possess these characters ; there exists, therefore, such a correspondence between the affection found (artificial animal tuberculosis) and human tuberculosis as can scarcely be imagined to be greater."

In his concluding remarks (p. 242), Orth says: "The possibility of contagion in animals once proved, there is the further conclusion also justified that contagion in man may be substantiated, . . . for I have already shown that tuberculosis in man and *Perlsucht* (bovine tuberculosis) must, according to their nature, be looked upon as identical diseases."

Now let us turn to the other authority mentioned by our author.

Professor O. Bollinger ("Ueber Impf- und Fütterungstuberculose," "Arch. f. exper. Path.," vol. i, p. 363), as one of the conclusions drawn from his own experiments, says : " Inoculation with tuberculous matter from man produces in the dog a genuine miliary tuberculosis of the pleura, lung, liver, and spleen ; usually, however, in carnivora no reaction at all, or only a local inconsiderable effect, is produced."

Apropos of a miliary tuberculosis of the peritonaeum,

liver, and lung, induced in a goat by combined inoculation and feeding with cheesy matter from the tuberculous lung of a cow, Bollinger (p. 360) says : "Corresponding with the resemblance of the macroscopic appearance to miliary tuberculosis of the human peritonæum, the microscopic examination of the miliary eruption yielded a perfectly analogous and identical structure."

Furthermore (p. 369): "As to infectiousness, tuberculous material from man and cow appear to be very similar." And again (p. 370), speaking generally, he says : " In agreement with the gross resemblance, the tubercle, produced by inoculation and by feeding, of the dog and the medium-sized domestic animals is also histologically identical with human tubercle."

A foot-note (p. 371) reads : " I have lately had an opportunity to observe a classic spontaneous miliary tuberculosis of the lung of a sheep, which in no manner differed from that of man."

The last words we will quote from Bollinger are found upon p. 371 : "If direct infection experiments upon man are at the present time wanting, nevertheless tuberculosis of the human intestines, which occurs secondarily in tuberculosis of the lungs, affords such an important analogue of the results of feeding experiments, and therewith so striking an example of self-infection through the swallowing of sputa, that one can, with great probability, admit the possibility of an infection by importation of the virus with the food."

So much for the real words and opinions of Orth and Bollinger as they themselves have expressed them in the papers referred to by our author. It might not be entirely out of place to remark that they also were written before the " bacilliary campaign" was inaugurated by Koch. We might modestly suggest that here is another opportunity for the author to display his unchallenged ingenuity.

Which will he prefer? To retract his assertion, to deny his own words, or to insist that Orth and Bollinger never wrote as they did?

The next standpoint, in the reverse order above mentioned, from which our author considers his first proposition is that of *inhalation experiments*. Still pursuing our original plan, we first quote the argument of our author:

"Tappeiner's induced inhalation tuberculosis of dogs (Virchow's 'Arch.,' lxxiv, 1878, and lxxxii, 1880), so much relied upon by Koch and others for the establishment of the mode of the spreading of phthisis, and partly of the bacillus doctrine itself, has been proved to be a fiction. Tappeiner, as so often quoted, subjected dogs to an atmosphere heavily charged with phthisical sputum, so that the dogs were nearly bathed in the latter (known to contain bacilli) for weeks. But, in spite of this, the animals grew fat, if anything, and, after the lapse of a certain time, acquired local pulmonary affections in the form of nodules, not likely to have been tubercular in their nature, of which only in one case were some observed in the liver and kidneys.

"The experiments of Schottelius (Virchow's 'Arch.,' lxxii, 1878, and xci, 1883), Wargunin and Rajewsky ('Vratsch,' No. 6, 1882), Weichselbaum ('Centralblatt,' No. 19, 1882), and others, and my own experiments also (to be reported subsequently), make Tappeiner's assertions perfectly untenable. Tappeiner's own account of his experiments, and the microscopical description of the structure of Tappeiner's 'tubercles' by Grawitz and Friedländer in Virchow's Institute, clearly indicate that he had nodular broncho-pneumonic foci, and not tubercles." ("Phila. Med. Times," April 5, 1884, p. 489, bottom of left and top of right column.)

Let us examine some of these statements and see how our author contributes to "the proof" that "Tappeiner's induced inhalation tuberculosis of dogs" is "a fiction." His objections to Tappeiner's experiments and conclusions appear to embrace four points: *a*, that the dogs were bathed for weeks in an atmosphere heavily charged with sputum; *b*, that they grew fat, and only in one instance showed disease of other organs than the lungs; *c*, that the disease produced was not tuberculosis, but broncho-pneumonia; *d*, that other experimenters (including himself) have produced broncho-pneumonia by inhalation of indifferent substances. Turning to the records, we see how truthfully our author has represented them.

Opposite the first point, *a*, we place the record of Tappeiner's experiments.

Tappeiner ("Ueber eine neue Methode Tuberculose zu erzeugen," in Virchow's "Arch.," Bd. 74, 1878, p. 394) says: "For all the experiments, the sputa were obtained from persons who were afflicted with tuberculous cavities of the lungs; from a teaspoonful to a tablespoonful thereof an emulsiform but transparent fluid was secured by dilution with 300 to 500 ccm. of water, which was conveyed to the experimenting-box by means of a steam atomizer attached to the outside. For experiments 1 to 8, the (inhalation) box—the dimensions of which were 1·12 metre in depth, ·82 metre in breadth, and ·86 metre in height—was open upon one side, being closed only by means of a grating, before which during the inhalation itself a waxed linen curtain was hung. Through a hole in the latter the stream from the atomizer entered the box. In experiments 1 to 4 the animals were subjected to inhalation twice daily for one hour at a time, and they remained the rest of the time in the box; in experiments 5 to 8 they were subjected to inhalation barely once daily, and the rest of the time were allowed to go free. In three further experiments the dogs were confined during the inhalation—given once daily, wit a very small amount of sputa, a teaspoonful in three days—in a roughly planked box of 12 cubic metres contents, in which, through numerous gaps in the walls, the air could pass in and out. As animals for experiment, dogs were exclusively selected, because they, according to Professor Bollinger, are only with exceeding rarity attacked with tuberculosis; in every case they could move freely about the inhalation-room, and, therefore, were not directly subjected to the stream from the atomizer."

Opposite the second point, *b*, we again place the records of Tappeiner's experiments.

Tappeiner (*loc. cit.*, p. 395) says: "The following are the results of the same eleven (experiments) performed according to the above-described methods, with a statement of age, weight of body at the beginning and at the end of the experiment, as also the duration of the latter:

"No. 1. Age, three years; weight at beginning, 6 kilogr., at end of experiment, 5·5 kilogr. Continuance of inhalations, 42 days. Autopsy: Pleural surface and parenchyma of both lungs thickly infiltrated with minute gray, semi-transparent nodules; in the kidneys and the liver, similar nodules in less number; the remaining organs were, at least by macroscopic examination, found uninvolved and normal.

"No. 2. Age, six months; weight at beginning, 8 kilogr., at end of experiment, 7 kilogr. Duration, 45 days. Same result as in previous case, only the nodules larger, confluent, less transparent, more yellowish.

"No. 3. Age, four years; weight at beginning, 20 kilogr.; duration of inhalations, 24 days; no loss of body-weight. Autopsy: Both lungs, as in previous cases, full of nodules, also visible discreet nodules in the kidneys.

"No. 4. Weight, 5 kilogr.; duration of inhalations, 25 days. Autopsy: Numerous miliary nodules in both lungs."

To these four observations may be added the following, found recorded in Tappeiner's second article—"Neue experimentelle Beiträge zur Inhalationstuberculose der Hunde." (*L. c.*, lxxxii, November, 1880.)

Tappeiner (p. 354) says: "For control (of experiments by inhalation of supposed indifferent matter) I allowed at the same time two other dogs, Nos. 1 and 2, in other inhalation-boxes (they were a metre square), to inhale phthisical sputa in the same minimal dose of half a gramme for both dogs."

It may be remarked, by way of explanation, that a half-gramme of phthisical sputa diluted with 100 grammes of water was used for inhalation once daily, fifteen minutes at a time, for ten days; at the end of the inhalation the dog remained in the inhalation-box four hours, and afterward was taken to the common dog-kennels of the veterinary school.

Of the results of these experiments Tappeiner speaks, p. 354 (*l. c.*), as follows: ". . . , while with Nos. 1 and 2 the spleen, as well as both lungs, was studded with numerous gray translucent nodules of the size of a millet-seed. . . . Virchow's second assistant, Dr. Grawitz, pronounced them undoubted true tubercles, both macroscopically and microscopically. . . . Dog No. 1 weighed 13 pounds at the beginning and 11¾ pounds at the end. No. 2 weighed 11½ pounds at the beginning and 10¼ pounds at the end."

At this time Tappeiner also conducted a series of experiments to discover, if possible, the length of the incubation period—i. e., the time at which the first visible outbreak of tubercle occurs after the beginning of the inhalations. Several examples of loss of body-weight are noted.

The following quotation is introduced, not so much to relate another instance of loss of body-weight after large quantities of phthisical sputa inhaled as to show that not only Tappeiner witnessed the results of his experiments, but often also other trustworthy and experienced men supervised them.

Further, Tappeiner says (p. 358) : " At my departure, on August 24th, I left in the dog-kennels of the Pathological Institute dogs Nos. 11 and 12. Both had . . . for fifteen days inhaled large doses of phthisical sputa, and hence must already have been infected with tubercle. One of these was killed by Dr. Grawitz on October 22, 1879, and the note of the autopsy was forwarded to me, on March 3d last, by Professor Virchow. It is in these words:

" Black poodle of Dr. Tappeiner, section Oct. 22, 1879. Small, extremely emaciated, hair almost entirely rubbed off on account of itch, black male poodle-dog. Belly greatly drawn in, covered with yellow adherent scabs.

" Heart firmly contracted, musculature dark-red, firm, valves unaltered.

" Left lung in places slightly adherent to the thoracic wall by a thin layer of fibrin ; those places appear cloudy, yellowish-gray upon the surface. The pleura itself is smooth as a mirror ; in it quite isolated numerous gray, translucent, miliary nodules are observable, the cut surface of which shows a very large number ; at the already mentioned places where the pleura was cloudy a broncho-pneumonic infiltration is found, reaching the size of a walnut. The right lung is smooth upon the surface ; it contains in the pulmonary pleura and in the pulmonary parenchyma numbers of gray miliary nodules ; in addition, small areas of flabby hepatization in the dependent portions. The spleen is slightly enlarged, hyperæmic, very dense ; follicles small, distinct ; no abnormality. The left kidney absent ! In its place a stellate scar, and a firm, whitish-gray lymph-gland the size of a hazel-nut. Right kidney correspondingly large, although not altered. Liver large, hyperæmic, in no way abnormal.

Intestine almost empty, mucous membrane pale, with abundant epithelial covering, indeed nothing.

"Diagnose: *Tubercula pulmonum et pleuræ.*—Pleuritis fibrinosa recens partialis sinistra. Broncho-pneumonia catarrhalis sinistra et pneumonia hypostatica dextra. Hyperplasia lienis chron. Defectus renis sin. Hypertrophia renis dextri. Hyperæmia hepatis."

Thus it is proved that, instead of only "one case" in which Tappeiner observed nodules in other organs besides the lungs, as our author asserts, *no less than five such cases are reported in the two articles referred to.* Furthermore, no less than three of the authors whom Dr. Formad quotes for other purposes have mentioned Tappeiner's experiments as having shown that nodules similar to those in the lungs are of frequent occurrence in the other organs when the duration of the experiment has lasted long enough. The cases quoted above also show that, when the experiments were performed under certain conditions, the animals lost weight and became emaciated instead of "grew fat." The whole series of experiments reported by Tappeiner also show that the greater the amount of sputa inhaled the greater is the tendency to loss of weight. It is true that two or three of Tappeiner's dogs increased in weight, but, as a rule, they were those which had inhaled the sputum in smaller quantity and with less frequency. Yet even these dogs were found to have serious disease of the lungs.

Opposed to the third point, *c*, we submit the following:

The records show that the series of experiments reported in Tappeiner's first article in many cases passed under the eyes of such men as Schweininger and Buhl, and that the series of experiments reported in his second paper were performed in Virchow's Pathological Institute, and under the supervision of such men as Grawitz, Friedländer, and sometimes even Virchow himself. And yet our author seems surprised and bewails himself that "Tappeiner has been so often quoted," and "so much relied upon by Koch and others for the establishment of the mode of the spreading of phthisis." Our author should give some weight to the fact that these experiments also were performed before the commencement of the "bacilliary campaign."

Whether or not Tappeiner, Schweininger, Buhl, Friedländer, Grawitz, and Virchow were capable of recognizing a genuine tubercle when they saw it, and were able to distinguish it from a broncho-pneumonic focus, I will not undertake to decide. But when I remember that these men were already renowned pathologists when our author received his medical diploma (which, by the way, was, if I remember rightly, the very year in which Tappeiner began his researches upon inhalation tuberculosis), I confess that, for myself, I am strongly inclined to prefer their authority, opinion, and testimony as to what they saw.

Indeed, as to the authority of our author, if we are to judge a man by his utterances, it may justly be said that there is ample ground for grave doubt if he possesses really any definite notions as to what a genuine tubercle is or appears to be. Furthermore, what is quoted below in justification of this doubt would seem also at the same time to warrant a suspicion that he is, in addition, still far at sea and greatly befogged in his knowledge of the nature of tuberculosis and pulmonary phthisis, in spite of those "long years of research he has almost exclusively devoted to this subject."

On page 340 of the "Phila. Med. Times," Feb. 9, 1884, left column, par. 8, appears the following language of the author: "No definite understanding concerning a disease can be arrived at unless some fixed conception of the anatomical characters and various expressions of the lesions of that disease is formed. Thus, as regards the question of tuberculosis and pulmonary phthisis, the matter would be much simpler if a general understanding could be arrived at as to the definition of tuberculosis and phthisis in its different anatomical manifestations. *The pivot of the question* is what to call a tubercle or a tubercular lesion." After floundering about in the vain search for "the pivot" among some floating thoughts more or less related to the matter discussed, our author seems at last to give up and in pure desperation grasp at the conclusion that "therefore it is impossible to define tuberculosis either by its anatomical peculiarity or by the pathogenic property of its nodes." (See *l. c.*, next page, par. 4, left column.)

Another specimen of our author's peculiar method of ratiocination frequently exemplified in his consideration of the ætiology of tuberculosis is subjoined.

After arguing against the conclusions of Tappeiner by misrepresenting his experiments, as has been already shown, our author proceeds to prove that neither general tuberculosis nor pulmonary tuberculosis can be produced in the dog by inhalation either of indifferent matter or of tuberculous matter, and to declare that his own experiments, which he *promises* shall be forthcoming at some future time, and "the experiments of Schottelius, Wargunin and Rajewsky, Weichselbaum, and others, make Tappeiner's assertions perfectly untenable."

As to the results of the author's own experiments upon this point, he will perhaps excuse us from attempting to estimate their weight in the settlement of this question until we are fully apprised of their character and details. We may be allowed, however, to offer the hope that, ere the future time of these promised experiments has arrived, the author will have found "the pivot of the question" and have learned "what to call a tubercle or a tubercular lesion."

As to the results of the other experimenters, they are, it is to be presumed, relied upon by the author to show that in dogs the products of the inhalations do not extend beyond the lungs and are nothing but nodules, more or less diffuse, of broncho-pneumonia. Very good. Now let us arrest ourselves for a moment and turn back again, to learn, if possible, what our author means by broncho-pneumonia.

On page 344 (l. c., Feb. 9, 1884, par. 2, right column) occurs the following sentence : "I class myself with those who regard all forms of pulmonary phthisis as tubercular." The succeeding paragraph commences with these words: "The lesions that are known as catarrhal pneumonia, *broncho-pneumonia*, pneumonic phthisis, cheesy pneumonia, tubercular phthisis, and fibroid phthisis, are all manifestations of the one disease."

This is the way in which the products of inhalation which Tappeiner has obtained are shown to be simple broncho-pneumonia and non-tubercular in their nature. In this manner would the conclusion of Tappeiner, that tuberculosis can be communicated even to the highly resistant dog, be "proved a fiction" and declared untenable by our author. It may be set down to his credit, however, that while he contests the conclusions of Tappeiner concerning dogs he generously admits "that the pulmonary tuberculosis may occasionally be produced in rabbits by these means." (Par. 2, right column, page 489, "Phila. Med. Times," April 5.)

In disposing of the fourth point, *d*, we might urge that all there is now left of it is that non-tubercular substances may sometimes produce, by inhalation, broncho-pneumonic foci in the dog, and leave this question just there until our

author has learned some method by which he can satisfy
even himself as to whether his broncho-pneumonic foci are
tubercular or not, and as to whether, by some fixed anatomical criterion, he knows how to distinguish between these
pneumonic foci and true tubercles.

But the citation by our author of the experiments of
Weichselbaum, as in harmony with those of Schottelins and
as a part of "the proof" which makes "Tappeiner's assertions perfectly untenable," gives occasion for their recital
here in order to show that Weichselbaum also has been
characteristically misrepresented.

As for Weichselbaum's own *deductions* from his experiments, the following shows not only that he is by no
means in accord with Schottelius (who maintains that there
is no difference between tuberculous matter and any indifferent material in the effect produced in dogs by inhalation),
but that, on the contrary, Weichselbaum is rather more in
agreement with Tappeiner.

Weichselbaum ("Inhalationstuberculose," "Centralblatt," xix, 1882, p. 340) says: "From the foregoing experiments it follows that not only tuberculous sputum, but
certain other organic substances, possess the faculty of exciting tubercle-like nodules, but that, in spite of this, contrary
to the results of the experiments of Schottelins, a difference
in the effect of the above-mentioned substances must be admitted—namely, in tuberculous sputum there is contained a
virus which, it matters not whether it enter the organism
in smaller or greater quantity, whether it be deposited in
the lungs or in the peritoneal cavity, calls forth, without
exception, nodules of tubercle-like structure in great number,
while other organic non-tubercular substances either can not
at all, or only under certain conditions, produce similar
nodules, and even then in small number."

But when the *experiments* of Weichselbaum are examined we find a still more complete agreement with and support of Tappeiner. These experiments comprised eleven upon dogs. Concerning the results of the latter, Weichsel-baum says: "In all these cases . . . there were found tubercles in the lungs and kidneys, in the former generally very numerous, in the latter only scattering." These experiments also comprised three with a watery emulsion of cheese. In two of these a stinking cheese (similar to Limburger) was used for inhalation. The two dogs died: the larger at the end of a month, after having received six inhalations; the latter at the end of five days, after two inhalations. The cause of death was gastro-enteritis, "while nothing of tubercle or anything similar was found." In the remaining one of these three cases, however, where a sort of cheese which did not smell badly was used for inhalation fifteen times in seventeen days, "twenty-four nodules the size of a millet-seed could be demonstrated in the lungs, and one submiliary nodule in each kidney, which showed a structure similar to that of the tubercles in the first-mentioned experiments."

"For control, two dogs were used; in the one dog, one Pravaz syringeful; in the other one, two syringefuls of the same cheesy emulsion used in the last experiment were injected into the peritoneal cavity; in neither case, however, were nodules found. The same negative result followed injection into two other dogs of tuberculous sputum which had been boiled for an hour, while unboiled tubercular sputum caused an abundant eruption of tubercles in the great omentum and mesentery.

"Furthermore, water-diluted pus from caries of a rib was employed for twelve inhalations within seven weeks, whereby a few scarce nodules, of a structure similar, however, to that of the tubercles of the first experiments, developed in the lungs.

"Finally a dog was made to inhale, sixteen times in twenty days, a watery emulsion of the spleen of an ox without the occurrence of nodules in a single organ."

Thus it is seen that of all Weichselbaum's experiments reported in the paper above mentioned, *only two* (viz., one of emulsion of non-stinking cheese, the other of pus from a carious rib) gave any warrant whatever for that author's deduction that " also certain other organic substances (besides tubercle sputum) possess the faculty of exciting tubercle-like nodules."

In criticism of these *two* experiments, it may be justly objected that, in the present state of our knowledge, it is impossible to know that the two organic substances used did not contain tubercle virus. Indeed, in the case of caries, according to a very widely accepted opinion among pathologists concerning the nature of that affection, it is highly probable that the carious pus used did contain tubercle virus. As for the use of cheese in experiments to prove that other substances than those which contain tubercle matter or virus are capable of producing tuberculosis, the objection is also obvious. It is founded upon the fact that there is a widespread belief among medical men, based upon the actual observations and experiments of known and respected authorities, that it is not only possible, but even probable, that milk from tuberculous cows may contain tuberculous matter. Who, then, can know that the particular piece of cheese used for experiment in this case did not come from a tuberculous cow ? What guarantee is there that in this emulsion some tubercular matter or virus was not suspended ? We are well aware that there are those, and probably our author is one, who may think such an objection a little strained. But it must be remembered—for it seems to be often forgotten—that *the one point* to be determined by all those experiments to produce artificial tuberculosis is now

not to learn if tuberculous matter can by inoculation produce tuberculosis, for this is now incontestable and is admitted by all respected investigators. The *one point* to be
fixed and determined is whether or not substances which
certainly contain no tubercular matter or virus can produce
the same disease. For the settlement of such a question it
is therefore necessary to be absolutely *assured* that the substances used for experiment can by no possibility contain or
be accompanied by tubercular matter or virus.

We have now reached, in our reverse order, the first
standpoint from which our author considers his first proposition : namely, that of *inoculation experiments.*

In the course of discussion the customary vein of the
author becomes diversified now and then by the introduc-
tion of a spirit of captiousness, of which the following may
be instanced as an example : Having again mooted the
question of the identity of human tuberculosis with that
produced experimentally in animals, our author proceeds
to annihilate " at one fell swoop " both Dr. Koch and his
knowledge of pathological histology by the declaration
that " it is hardly within the province of the mycologist to
teach us what is tubercle and what is not." (" Phila. Med.
Times," April 5, p. 488, right column, par. 6.)

There are one or two statements in this portion of our
author's consideration of his first proposition which deserve
to be noticed here.

Apropos of the discussion of Koch's inoculation experiments and statements concerning the relation of the tubercle
bacillus to tubercular lesions, Dr. Formad seems to deny
that there is any positive and reliable means of distinguishing the tubercle bacillus from other bacilli, for he objects to
attempting to diagnosticate, by means of the bacillus, lesions
which appear to be tuberculous and which contain tubercle
bacilli, " because similar bacilli may be found in the lesions

of various processes resulting in cheesy products (see bacillus chapter)." (*L. c.*, par. 4.) Here the author is not only in conflict with nearly the whole world of investigators, but, as usual, is also in combat with himself.

Referring, as directed, to the bacillus chapter, we find that the author, after a long review of the published observations upon the occurrence of the tubercle bacillus in the sputum and elsewhere, has summarized his views concerning the peculiar diagnostic characteristics of the tubercle bacillus as follows: " From our present knowledge of the occurrence of Koch's bacillus in the sputum, we must therefore conclude: 1. That the presence of bacilli is a valuable *diagnostic sign* of tubercular disease of the lung." (*L. c.*, March 22, 1884, p. 452, right column, pars. 2 and 3.)

Further: " To detect bacilli is a very simple matter, although by far not so easy as to prepare a specimen of urine and to find the all-important tube-casts." (*L. c.*, p. 445, right column, foot-note.)

And again we quote the author's words: " It (the bacillus tuberculosis) is concomitant with most tubercular lesions. *It is diagnostic of tubercular change.* It is, on account of its irritant properties, one of the causes of tuberculosis." (*L. c.*, Feb. 9, 1884, p. 338, left column, par. 3.)

Finally, we quote from his conclusions at the end of his paper: " From the above analysis of the bacillus question and of the ætiology of tuberculosis the conclusions follow: 1. That the bacillus of Koch is a valuable diagnostic sign of tubercular disease." (*L. c.*, April 5, 1884, p. 497, left column, par. 5.)

We might rest here, but at the latter end of the chapter referred to there are two important affirmations and arguments built thereon, which should not pass unchallenged. It is asserted that " bacilli, not distinguishable from tubercle

2

bacilli, are met with in lupus and leprosy." (*L. c.*, p. 453, right column, par. 6.)

When our author says, in support of his statement as to lepra, that "the bacillus of leprosy, in specimens which I had the opportunity to examine, appears to me also perfectly identical with the small forms of tubercle bacilli ; although the lepra bacillus may, perhaps, look more sharp-pointed to the eyes of others, and may fail to take the brown stain" (*l. c.*, par. 7), he states his only argument in support of this declaration as to the lepra bacillus, and at the same time utters a sentence of which the last half is antagonistic to the first half. Each member of this sentence is assailable, and is, moreover, positively misleading. In estimating the value of our author's testimony concerning the perfect identity of the tubercle and lepra bacilli, it is important that something be known of the character and extent of the examination of the lepra bacillus he claims to have had the opportunity to make. It is not impertinent to this matter to inquire why the statement of *the whole truth* concerning this examination of the lepra bacilli has not been made. Is it possible that the author fails to appreciate the right of the reader to know the qualifying facts, that the few specimens of lepra bacillus examined by him were permanent preparations in the possession of others, that he could not personally vouch for their genuineness, and that he knew nothing of the methods by which they had been stained ? Take the last member of this somewhat characteristic sentence ; it also contains a characteristically inaccurate statement of fact. The concluding phrase thereof constitutes an expression of fact directly opposite to the truth, and reads as follows: "and (although the *lepra bacillus*) may fail to take the brown stain." The truth relative to the distinguishing micro-chemical reaction of the two bacilli in question is accurately stated thus: "and (although the *tubercle bacillus*) may fail to take the brown stain," while, on the contrary, the lepra bacillus may readily absorb it. (Refer to Koch's views on this point, quoted in the first part of this article.)

As to the author's objection that "bacilli, not distinguishable from tubercle bacilli, are met with in lupus," the answer is patent. First, let it be absolutely demonstrated that in none of the forms of lupus have we to do with a disease mainly local in the skin, but which in its nature is really tuberculous. Until that be done it is not only futile, but it is a waste of energy, to urge that the few scattering bacilli which very rarely have been found in the lesions of lupus are not genuine tubercle bacilli.

The author demands absolute proof at the hands of Koch and others who hold to the belief of the specific nature of tuberculosis. Let him also, as well as others, rely upon and advance evidence which is absolutely unassailable. There are now, and for many years have been, not only distinguished dermatologists, but also equally respectable pathologists and clinicians, who have been, and are, inclined to recognize lupus, scrofulosis, and tuberculosis as mere variations, due more or less to local conditions of the action of one and the same morbific cause.

There is one more argument used by our author in the discussion of his first proposition from the standpoint of inoculation experiments which deserves to be considered.

In objecting to the use of the tubercle bacilli as one of the means of diagnosticating a lesion in which they are imbedded, he says : "Besides, there are many spontaneous and artificially induced tubercular lesions in which bacilli could not be found." (*L. c.,* p. 488, right column, par. 4.)

Among the many considerations which weaken the force of this argument, the following may be mentioned. As customary, we state some of them by reproducing the author's own words.

In discussing the fact that tubercle bacilli are not always discovered even in phthisical sputum, in which, as every one knows, they are far more easily demonstrated than elsewhere, the author thus expresses himself: "The examination of sputum may thus in doubtful cases be quite misleading; for, if in any given case bacilli are not found, it should be taken into consideration . . . that the examiner may fail occasionally in any case to succeed in preparing a successful preparation of stained bacilli." (*L. c.*, March 22, p. 452, left column, par. 5.) How much the more is this true of the preparation and examination of sections of tissue in which the tubercle bacilli are imbedded, the experience of all workers shows. If occasional failure to see the bacilli in phthisical sputum does not invalidate the use of the tubercle bacillus as a *diagnostic sign* of the existence of the tuberculous process, certainly no man can consistently deny that the same tubercle bacilli may be used also as one of the means of diagnosis of tubercular products which contain them, simply because in such products the examiner may occasionally fail to see the tubercle bacillus.

As having a direct and important bearing upon the value of any deductions which the author may be disposed to draw from his failures to discover tubercle bacilli in certain tubercular lesions, it may be proper to examine his following statement: "A magnifying power of 400 diameters is nearly always sufficient to detect stained tubercle bacilli. In fact, we found that when we failed to find bacilli with a good $\frac{1}{8}$ objective, neither our $\frac{1}{12}$ Zeiss oil-immersion lens nor the Abbée condenser would reveal any when used (as we always do) for control." (*L. c.*, March 22, p. 446, left column, par. 1.)

Those who have been actual workers both with the high

and the low power mentioned, both with and without the use of the Abbée condenser, can not but be struck with the absurdity of such a statement coming from one who would pretend to speak with authority about the presence or absence in the tissues of an object in every way so difficult of demonstration as is the tubercle bacillus when *in situ naturali* in small numbers. It is, of course, possible, as every one knows, to use a magnifying power of 400 diameters, such as can be obtained by a good ⅛ objective and a strong eye-piece, without the Abbée condenser, in the common clinical examination of spnta. But when any one says or ventures to intimate that he can, with even an approach to certainty, recognize isolated tubercle bacilli, in sections of tissues prepared for the microscope, without the aid of an Abbée or a similar sub-stage condenser and the use also of an excellent objective of very high power, neither his negative observations nor the conclusions drawn from them will be credited by those who are practically familiar with both methods of examination. Yet our author has drawn most positive and important conclusions, not only from his failure to find any bacilli whatever in some tuberculous tissues, but also from his reported failure, in some instances, to find them in satisfactory numbers. From my own experience in the use of high and medium powers, with and without the aid of the Abbée sub-stage condenser, in the search for bacilli in the tissues; from the fact that I have never seen an Abbée condenser in use in the Pathological Laboratory of the University, or known such an instrument to be in common use there; as well as for other reasons not necessary to mention, I am free to confess that, for myself, I decline to accept these reported failures of the author as in any degree conclusive that in these very cases tubercle bacilli were not to be shown by proper methods and means of demonstration.

The author refers (*l. c.*, p. 446, right column, par. 4) to our joint examination, three months after our return to America, of a most beautiful and perfectly pure tubercle-bacillus culture with which Dr. Koch had kindly presented me. The flat salt-dish within which this culture was inclosed was opened for the first time in the University Laboratory, and some blood serum already prepared by the author was inoculated with a few of the scales. At the same time, several preparations of this culture were at once made for microscopic examination, the staining fluids belonging to the laboratory being employed. At the first examination of these preparations, made with the $\frac{1}{8}$ lens commonly used in that laboratory for the examination of sputum and other materials in searching for tubercle bacilli, it was impossible for either of us to determine whether we had under the microscope a pure culture or not. There were many rod-forms so short as to be practically unrecognizable as bacilli by means of that magnification and lens. Yet, when examined by the $\frac{1}{12}$ Zeiss homogeneous immersion, there was no longer any possible doubt that every bacterium visible was a characteristic tubercle bacillus.

Furthermore, the author refers, in his consideration of tuberculosis of the serous cavities, to four cases of " primary peritoneal and pleuritic tuberculosis " which have come under his observation during the last eighteen months, and takes occasion to state that " no bacilli could be discovered, even after repeated and careful search, in any of the lesions." (*L. c.*, Feb. 23d, p. 381, left column, par. 4.)

Inasmuch as the author has done me the honor of mention in the paragraph preceding the last quoted, I may, perhaps, be excused if I offer here some testimony of my own bearing more or less directly upon the last-mentioned negative observations of the author. Gross specimens of the abdominal lesions from one of several cases of primary tuberculosis of the peritonæum reported by our author as containing no tubercle bacilli recently came into my hands. A few isolated, small miliary nodules upon the peritonæum were snipped off carefully and quickly ground to a pulp (in a new agate mortar which had never been used and had been sterilized), with the addition of a few drops of water which contained no tubercle bacilli. Some of this pulp was spread in a thin film upon a glass cover, in the manner of preparing sputum; this film was stained in the usual manner, after Ehrlich's method as employed by Koch. It was examined by me with a fine $\frac{1}{10}$ immersion objective, with the aid of an Abbée condenser, and was found to contain a small number of tubercle bacilli which were characteristically, although somewhat faintly, stained.

But there is yet a more serious reason for declining to accept as conclusive the assertions of our author concerning his own experience with the tubercle bacillus. Loyalty to the cause of pathological research, as well as the gravity of the subject discussed, demands that it should be stated. I refer to an evident disposition on the part of the author, frequently manifested even in print, to magnify and misrepresent the observations which he undertakes to report.

Perhaps one of the most glaring examples of such a disposition to be found in the second paper of the author is the following:

Speaking of the well-known experiments of Ziegler, in which plates of glass were inserted into the subcutaneous tissue, our author has thus printed his observations:

"Ziegler very properly declared the latter product to be tubercle tissue. I have had, and have at present, ample opportunity to corroborate the accuracy of these observations. Ziegler's experiments were repeated in the Pathological Laboratory of the University of Pennsylvania by Hammer, and at present are being carried on by Woodnut.

By these experiments, made, with slight modification, after the method of Ziegler, under varying conditions and upon various animals, it was shown that the granulation tissue gradually gave origin to tubercle nodules. Furthermore, these experiments showed that *the tubercle nodules* and *cheesy changes ensue without* the *action of bacilli*, as the latter were found not to be present where proper care was taken during the execution of the experiment to exclude them." (*L. c.*, Feb. 9, 1884, p. 342, left column, par. 1.)

With respect to the above-cited experiments of Wood-nut we have nothing to say except that he was during that time not yet even a graduate in medicine, but only a young medical student, presumably more or less skillful, or more or less careless and inexperienced. But what of importance there is to be said relates to the above-mentioned experiments of Hammer. It is simply this: that it is a matter of common knowledge among the university people that Hammer's experiments and observations on this subject were both begun and completed long before Koch's first announcement of the existence of the tubercle bacillus, and that no one was better aware of that fact than the author when that statement was deliberately written for publication.

We have now reached the end of our criticism of the manner in which our author has considered his first proposition in the one page and a half which he devotes to that matter. After having thoroughly and carefully examined the two papers published by our author on the ætiology of tuberculosis, we are convinced that there is ample justification for the declaration that it is difficult to find, from one end to the other of these two papers, any two consecutive pages of print which are not equally defaced with similar faults. But we stated at the outset that we possessed neither the time nor the inclination to touch upon all that appeared to us objectionable.

We shall end this communication by a brief examination of the sort of logic by which our author concludes that the tissues of the scrofulous possess a specific anatomical structure, and that the ætiology of tuberculosis is thereby explained.

During the discussion, before the Philadelphia Medical Society, of the author's last paper on tuberculosis, we took occasion to state some objections to the acceptance of his hypothesis concerning the lymph-spaces of the so-called scrofulous. (See "New York Medical Journal," June 14, 1884.) The reasons then urged will not be repeated here, but we will limit ourselves at this time to pointing out some statements of our author which, to our mind, are fatal to his hypothesis. Before quoting these statements, however, it may be well to call attention to a very serious gap in the connection between this alleged peculiarity of the lymph-spaces of the connective tissue, which our author has as yet made no effort to fill up by direct observations. We allude to the fact that, while the author claims to have demonstrated, in the scrofulous, the existence of a peculiar narrowing of the lymph-spaces in the connective tissue of the skin, of the subcutaneous tissue, and possibly also of the intermuscular connective tissue, he has entirely neglected the lymph-spaces of the lungs, the liver, the spleen, the kidneys, and the bones. One does not need to be a pathologist to be cognizant of the extreme rarity of tuberculosis of the skin, and of the extreme frequency of tuberculosis of those very organs whose lymph-spaces have not yet been studied by our author.

We now proceed to indicate the difficulties which our author has, according to his custom, placed in his own way.

After having reaffirmed his former assertions and conclusions concerning the lymph-spaces in the connective tissue of the scrofulous, our author thus delivers himself:

" For details of my researches in this direction, I must refer
to my first paper upon this subject." (*L. c.*, Feb. 23d, p.
378, right column, par. 2.)

Referring to that first paper we find the following :
" Comparing a large number of sections taken from corresponding parts of the bodies of rabbits and cats, it is also
distinctly seen that the lymph-spaces are on the average de-
cidedly narrower and fewer in the rabbit than in the cat.
*The perivascular spaces are, however, equally free and similar
in both.*" (*L. c.*, Nov. 18, 1882, p. 113, right column,
par. 1.)

Returning again to the second paper, we find the subjoined statements as to how and where tubercle originates :
" *Primary* tubercle occurs as a mere infiltration of lymphoid
cells *in the adventitia of blood-vessels* (a term synonymous
with perivascular lymph-spaces), or as small nodular masses
of lymphoid infiltration around blood-vessels or ducts of
any kind; or tubercle tissue may organize within blood-
vessels and various ducts." (*L. c.*, Feb. 9, 1884, p. 342,
left column, par. 3.)

Of the occurrence of *secondary* tubercles thus he has
written : "These seem to lie *in the perivascular lymph-
spaces*, and are probably distributed throughout the body
mainly by means of these lymph-channels of the blood-
vessel walls. Tubercles do not occur in avascular tissues."
(*L. c.*, right column, par. 3.)

We confess our utter inability to see how the occurrence
of tubercles in locations where the lymph-channels are admitted by our author to be "equally free and similar in
both" the so-called scrofulous and the non-scrofulous animals can have anything whatever to do with peculiarities
which may possibly exist in other places. Perhaps the
inventive ingenuity and surpassing skill in engineering
of our author may enable him to remove some of these

self-imposed obstacles, or to vault over them or brush them aside. For ourself, in their present shape they appear insurmountable.

After the author, by revising a statement here, retracting one there, and by filling up that gap and this, shall have succeeded in bringing his *hypothesis* into complete accord with facts and common clinical experience, it will then remain for him to show that the claimed peculiarity of the lymph-spaces of certain men does not have quite as much to do with the ætiology of the lesions of syphilis as with those of tuberculosis. (See Cornil on "Syphilis," American edition, by Simes and White, pp. 24, 25, Philadelphia, 1882, and "Clinical Lectures on the Physiological Pathology and Treatment of Syphilis," by Fessenden N. Otis, M. D., New York, 1881.)

In the foregoing incomplete criticism of the author's printed assertions concerning the ætiology of tuberculosis we have, as far as possible, purposely avoided a general discussion of the latter subject, and have intentionally adhered closely to the text of his own papers, and to that of a few of those to whom he refers.

Our sole object has been an endeavor to discover, if possible, what degree of reliance, as a witness whose testimony is in conflict with that of Koch and others, can safely be placed upon the observations and conclusions which our author has already published as his contribution to the knowledge of the ætiology of tuberculosis.

•We readily confess ourself greatly at a loss for an adequate standard by which to measure that degree of reliance which they intrinsically deserve, for, so far as our reading of the extensive, extremely varied, and conflicting literature of this much-discussed subject has gone, it appears that these contributions of our author stand unique and without comparison. Perhaps the only practical gauge, after all, is that which each one will apply for himself. We leave the application to be made.

Finally, we are also somewhat at a loss for proper terms with which to conclude. Since, however, in the preceding pages we have so frequently turned the words of the author against himself, it may not be improper to continue this course to the end.

"Thus, above all, *negative evidence* must be carefully inquired into, not by relying upon the crippled and sometimes misrepresenting and meager quotations of some compiling writers, but by submitting the original communications of the authors and experimenters to a careful perusal." (*L. c.*, April 5, 1884, p. 494, left column, par. 3.)

We have pointed out striking "instances of the way in which an experimenter with preconceived and peculiar ideas upon a subject may unconsciously be misled in forming conclusions from his own experiments." (*L. c.*, April 5, 1884, p. 495, left column, par. 2.) We have at the same time indicated how our own author, as well as "some of the younger German pathologists and a few of the prominent English surgeons, under the influence of the bacillus craze, will make in publications assertions entirely unwarrantable." (*L. c.*, Feb. 9, 1884, right column, par. 1.)

"All of the above statements are made by a scientific man and pathologist, and offered as broad facts in full earnest. I only have to say that here evidently observation is substituted by imagination and mere speculation; and all this is done for the sake of the convenience in explaining a disease by pretty hypotheses." (*L. c.*, par. 2.)

nounced officially the death of Dr. Hunt, the physician and journalist. He came to Buffalo in 1853, and was associated with Dr. Austin Flint in the editorship of the "Buffalo Medical and Surgical Journal." He was for several years professor of anatomy in the University of Buffalo, a position he filled admirably. But as a medical journalist we had never had his superior. He was for several years the secretary of the Buffalo Medical and Surgical Association, and his reports were models of excellence; indeed, so facile was his pen that many gentlemen scarcely recognized their own remarks. He finally left the ranks of the profession to become a regular journalist of the daily press, yet not before he had shown marked ability and skill as a surgeon in the army during the war. After an absence of twenty years, Buffalo was still so dear to him that he desired to be buried here. Dr. Rochester moved that our regret at his loss to us and to the profession be placed upon the minutes of this meeting as a tribute to his memory and a mark of respect to his family. Carried.

Dr. Wyckoff, Dr. Gay, and Dr. Strong severally eulogized his intellect and talents, and recalled pleasant associations with him of a score of years ago.

Aneurysm of the Aorta.—Dr. Hayd presented an aneurysm of the aorta removed at an autopsy this morning. It was interesting from the fact that the young man was a diver, and was accustomed to remain under water ten hours or more with a bell, causing a great strain upon his system. This led undoubtedly to a chronic endarteritis deformans, with dilatation and sacculation of the first part of the aortic arch. The sac contained several stratified deposits of fibrin. There were a few végétations globuleuses upon the aortic and mitral valves. The aortic valve was thickened.

Gall-stones in the Liver.—Dr. Hartwig reported a case observed by him for some months, in which he had made a diagnosis of hepatic abscess, while other physicians had called it hepatitis. The patient, who had been suffering for two years, had little or no pain over the liver, but there was enlargement of the organ. He introduced a hypodermic needle into the liver in two places without finding pus. On post-mortem examination, there were found several thousand small abscesses in the liver, each containing a calculus. The gall-bladder also was completely filled with gall-stones. Some suppuration had taken place in one of the wounds made by the needle, and he believed puncture of the liver with a needle to be dangerous, especially on account of the great vessels at the hilum.

Dr. Cronyn observed that piercing an artery with a needle was not at all hazardous, for the tissue about the puncture contracted at once.

Dr. Hartwig said he was loth to believe this unless it was proved by actual experiment upon animals.

The Proposed Division of the Association into Sections.—Dr. Coakley, chairman of the committee appointed at the last meeting to report upon this subject, stated that a circular had been sent to each of the hundred members of the association, requesting the enrollment of his name in one of the four proposed sections, viz., medicine, surgery, obstetrics, and pharmacy. He had received only fifteen responses, and several of the writers were opposed to the division.

In a short discussion which followed the remarks of the chairman, Dr. Treszline, Dr. Cronyn, and Dr. Hartwig expressed themselves against the change, believing the association was as yet too small, the specialists in it too few, and, further, that it was wisest to make ourselves specialists in general medicine.

A motion of Dr. Andrews that the committee be continued was carried and the meeting adjourned.

FREDERICK PETERSON, M. D., Secretary.

PHILADELPHIA COUNTY MEDICAL SOCIETY.

Meeting of May 21, 1884.

The Bacillus Theory of Tuberculosis.—[After the reading of an interesting paper on "The Type of Typhoid Fever Prevalent Last Winter, with Pathological Specimens," by Dr. George W. Vogler; and a "Warning to Medical Practitioners in regard to the Use of Jequirity," by Dr. M. Landesberg—both of which we are obliged to omit on account of lack of space—the society proceeded to the discussion of Dr. Formad's paper on "Tuberculosis," as follows:]

Dr. Shakespeare regretted his inability to be present to open the discussion in accordance with the request of the President, and thanked the society for this opportunity of expressing his views. He had been much interested by the opinions and by the review of the status of the tuberculosis question presented by the author. There were, however, very many points assumed as demonstrated, and positive statements advanced in the elaboration of Dr. Formad's paper, which Dr. Shakespeare believed to be without sufficient foundation. But he would not, at this time, enter into a general criticism. He preferred to await the detailed observations which the author promised should be forthcoming in support of the many statements and conclusions he had thought proper to announce in advance. He intended to limit his remarks to-night to some differences between himself and the author as to statements made by the latter concerning a recent visit to Koch's laboratory. Dr. Shakespeare also had been in Berlin last summer, and had then enjoyed the privilege for about a month of working under Koch and his assistants during six or seven hours daily.

1st. The author had declared, in terms far less equivocal than those printed, that Koch's policy was to hinder or prevent strangers who visited the Gesundheitsamt from retracing his now famous experiments upon tuberculosis, and stated that no one had ever been permitted to inquire into the infectiveness or parasitic nature of tuberculosis, save one man.

2d. The author had further announced that Koch had so far modified his views that he now admitted that neither the form, size, and aspect of the tubercle bacillus, nor its want of individual motion, nor its peculiar behavior toward staining fluids, distinguished it from many other bacilli.

Dr. Shakespeare regarded these statements as misrepresentations of Koch's animus, as well as of his present opinions. He felt impelled to thus publicly express himself, because perhaps every member present had known of his late visit to the Kaiserlichen Gesundheitsamt. To be silent under these circumstances would constitute a tacit assent to these declarations—a false position in which he was unwilling to be placed. Moreover, the grave importance of this whole question; the presumed desire of this learned society to be possessed of all the evidence bearing upon every phase of it; justice to the fairness, honesty, and consistency of the distinguished author of the bacillus theory of tuberculosis, whether it were true or false, forced him to express now his dissent from the foregoing declarations of his friend.

Previous to the announcement of the discovery of the "tubercle bacillus" he had been most favorably impressed by the exactness and completeness of Koch's labors in the final establishment of the parasitic nature of anthrax (French, charbon; German, Milzbrand; English, splenic fever), as also by the evident caution and reliability of that investigator. This had prepared him to begin the examination of the grounds of Koch's startling claims regarding the nature of tuberculosis with no small degree of respect for their author. At that time he had no definite views concerning the cause, infectiousness, or contagiousness of tuberculosis. Certainly he did not commence this examination with a mind wholly preoccupied by a theory of his own which he thought to be in conflict with that of Koch.

He had not gone to Berlin for the purpose of discovering there the truth or falsity of the claims for the "tubercle bacillus." On the contrary, recognizing the growing importance of research among the various forms of bacteria as possible causes or modifiers of pathological processes, and having personally experienced much trouble in prosecuting such studies while following described methods, and, through his intimate relations with the University of Pennsylvania, having known of similar difficulties in the pathological laboratory of that school of medicine, he had at length determined to obtain, if possible, ocular demonstration of Koch's classic methods of isolation, culture, and study of minute organisms, and had become one of "the pilgrims" to that Mecca toward which Dr. Formad himself had directed his steps only a few weeks before.

Arrived in Berlin, he had been most cordially welcomed at the Gesundheitsamt by Dr. Koch. Koch and his corps of accomplished co-laborers, and every possible facility for furthering the object of his visit had been most willingly and courteously tendered during the whole of his stay, though doubtless at the cost of much inconvenience, for, besides work upon important investigations, active preparations for the departure of the cholera expedition to Egypt were then in progress. He could say that he had never spent a month with more pleasure or profit. While it had not been his desire to give especial attention to the "Bacillus tuberculosis" more than to the Bacillus anthracis and to other bacteria, yet, so far as his wish extended, and the limited time at his disposal served, in his practical work the "Bacillus tuberculosis" was not neglected.

He felt impelled to say, in the most emphatic and unmistakable language which he could use, that he himself was not only readily permitted to go as far as he wished in the investigation of the tubercle bacillus, but, furthermore, on no single occasion did he meet with any hindrance whatever, or perceive the slightest indication of a desire on the part of Koch to prevent the retracing of his experiments upon that subject. He had heard of no one having met such a difficulty there other than Dr. Formad. The only person who, previous to the presentation of the paper under discussion, had to his knowledge published an account of personal work done upon tuberculosis in Koch's laboratory was Watson Cheyne, of England, whose report amply testified to Koch's willingness to have his experiments examined. Dr. Formad, in his communication as printed, excepted this work of Watson Cheyne, perhaps wisely, for he several times quoted it for other purposes.

If Dr. Formad, during the three or four days of his attendance at Koch's laboratory, did not experience an enthusiastic reception, and, as he intimated, was not permitted to experiment upon the pathogenic qualities of the tubercle bacillus, he might far more reasonably have attributed this coldness to an irritation naturally produced by his published remarks in which Koch had been accused of unscientific work, and the insinuation been offered that the researches made at the Imperial Health Office had been unduly influenced by Kaiser Wilhelm, than have assumed from his reception that Koch habitually objected to have any one look into the genuineness and reliability of his work upon tuberculosis. Indeed, the simple fact of his admission at all, under the circumstances, could fairly have been regarded as evidence of Koch's willingness to open his laboratory even to an opponent whom he regarded as unfair. The Gesundheitsamt was a department of the German Government. Koch and his chief assistants were officers of the German army or navy. They were all intensely loyal to their Emperor.

They believed that Dr. Formad had purposely and unjustifiably stepped outside the proper sphere of a purely scientific communication to publish a reflection insulting to them and their Kaiser.

Before dismissing this indirect attack upon the reliability of Koch's published observations upon tuberculosis, Dr. Shakespeare took this opportunity to say that his personal observation of Koch, as well as a careful examination of his publications, had led him to the conviction that the whole medical fraternity did not possess a more painstaking, capable, cautious, thoroughly honest and reliable investigator of the causes of disease than the distinguished discoverer of the tubercle bacillus. He would speak in similar terms of those of the corps of official co-laborers at the Gesundheitsamt with whom he had come in contact sufficiently often to form an opinion.

The second statement above mentioned, namely, that Koch had now essentially modified his views concerning the characteristics of the tubercle bacillus, was next examined. Dr. Shakespeare could only say that Dr. Formad's extraordinary announcement was the first and the only information upon this point which he had received. Certainly he had heard nothing and seen nothing while at the Gesundheitsamt which could in any manner confirm such a statement. It is true that, while at Berlin, the author had related to him his interview with Koch, and had said that the latter had been far less dogmatic than he had expected, mentioning among other things a little friendly controversy concerning their opposite views in which Koch had seemed quite willing to admit the possibility that under favorable circumstances the tubercle bacillus might develop a flagellum at its extremity, and thus become endowed with individual motion (Dr. Formad had professed to have seen this motion), and had appeared quite willing to admit also the possibility that in the course of time it might be discovered that other bacteria would react toward staining fluids in a manner identical to the reaction of the tubercle bacillus. But an admission that certain things might be possible and a statement, based upon present knowledge and experience, that they did exist, or even were probable, were quite different matters. During Dr. Shakespeare's work upon the tubercle bacillus in Koch's laboratory, which was after the termination of the short visit of Dr. Formad, he was taught to differentiate the tubercle bacillus from all other bacilli by means of its characteristic reaction, now well known, toward certain staining agents, no less than by its peculiar size and shape, as seen under high magnifying powers (Zeiss's $\frac{1}{12}$ was generally used for this purpose). The statement that the author of the bacillus theory of tuberculosis had practically withdrawn his claim that there was something characteristic in the staining of the tubercle bacillus and in its morphology which distinguished it from other bacilli was the more astonishing and incredible because of the fact that, besides the existence of overwhelming testimony from all quarters of the globe in confirmation of this original claim, even Dr. Formad, however persistently in print he might assail this claim of peculiarity, was himself in the habit of differentiating this minute organism from all other known bacilli for purposes of diagnosis and of demonstration to his pupils by means of this self same characteristic coloring and morphology.

Although it had not originally been his intention to discuss them this evening, Dr. Shakespeare briefly considered Dr. Formad's claims of discovery of the ætiology of tuberculosis as set forth in his two papers. This author had been among the first to controvert Koch's theory of tuberculosis. Somewhat more than a year ago he made the first announcement of his views. In this communication the author advanced a theory of his own, which he believed to be opposed to that of Koch. He maintained that there was no necessity for the action of a specific agent in the production of tuberculosis, and that therefore such a specific agent could have no rational existence. This position was, in the main, based upon his belief in the discovery of an anatomical peculiarity of those animals known to be especially prone to tuberculosis. This peculiarity he thought to consist essentially in a narrowing of the connective tissue lymph-spaces in certain animals—the scrofulous—and to be either hereditary or acquired. He maintained that the inflammatory process in such animals, whatever the exciting cause, was necessarily tuberculous.

On the occasion of the presentation of his first paper, Dr. Formad had undertaken to demonstrate this reputed anatomical peculiarity by the exhibition, under the microscope, of a number of anatomical preparations. At that time Dr. Shakespeare had regarded that demonstration as far from satisfactory or conclusive. In the first place, no single section showed lymph-spaces. In the second place, the method of preparation followed (that for ordinary histological examination—hardening in alcohol, cutting thin sections, staining these with carmine, mounting them for examination in Canada balsam) naturally was not capable of demonstrating lymph-spaces; not one silver or gold preparation was exhibited. Indeed, this common and satisfactory method of studying lymph-spaces had apparently not even been resorted to, for it was to be presumed that the most positive and demonstrative specimens in the possession of the author were those selected for exhibition. It was true that some of the sections under the microscope showed a cellular hyperplasia of the connective tissue—an appearance by no means new to the scientific world. And this was the sole evidence presented in support of a reputed discovery concerning an important anatomical peculiarity of the lymph-spaces of so-called scrofulous animals, upon which an exclusive theory of the ætiology of tuberculosis had been erected by the author and claimed to be demonstrated.

Recognizing the importance of that reputed discovery, this learned society had at once appointed a committee, consisting of its most experienced microscopists, to examine anatomical preparations which Dr. Formad should lay before it in proof of his announced discovery. Nearly eighteen months had since elapsed, and yet, during all that time, not one preparation had been submitted for examination by that committee.

In the paper at present under discussion, the author had complacently referred for proof of his so-called discovery to the evidence brought forward in his first paper, and supplemented this by promising with apparent self-satisfaction the future publication of corroborative observations by some independent investigators. Other criticisms might justly be urged, but, in view of the foregoing facts alone, Dr. Shakespeare believed himself sufficiently warranted in contending that the basis of Dr. Formad's opinion concerning the ætiology of tuberculosis had not been established, and also in suggesting that, instead of that opinion being referred to as a "theory" against the theory of Koch, it was scarcely yet entitled to be dignified by the name of hypothesis.

Furthermore, even admitting that this hypothesis concerning the anatomy of the lymph-spaces of the so-called scrofulous animals were, by the most indisputable evidence, demonstrated beyond the possibility of doubt, it still contained absolutely nothing which by itself either necessarily supported the conclusion of Dr. Formad regarding the non-specificity and non-infectiousness of tuberculosis, or antagonized the claim of Koch for the specific pathogenic qualities of his tubercle bacillus. When, if ever, this hypothesis became a fixed and determined fact, we should be placed only one step nearer a correct understanding of the ætiology of tuberculosis. The reason of that peculiar predisposition which certain animals were known to show toward tuberculosis might then have been satisfactorily ex-

plained. But what the exciting cause of that peculiar malady might be was an entirely different question. Whatever this might be, it could be readily understood that its power of destruction would naturally be favored by such an anatomical peculiarity. Such an "anatomical peculiarity," if it really existed at all, could be easily turned to the support of the bacillus theory. The claim of Koch was not that the tubercle bacillus was endowed with pathogenic qualities which under any and all circumstances were capable of exciting tuberculosis. He himself declared that for the calling forth of those powers a suitable soil and conditions favorable to growth and propagation were essential.

Finally, Dr. Shakespeare thought it proper to define his own position with regard to the ætiology of tuberculosis. He wished it to be distinctly understood that it was not from the standpoint of a follower of Koch, who accepted all of that investigator's conclusions, that he had offered the criticisms which he had made. In the consideration of such a grave question as the one then confronting him, he regarded it as obligatory to exact the same degree of rigid proof from friend as from foe, whether advanced on the side of popular opinion or against it. He therefore had not hesitated to express objections to the opinions and statements advanced by his friend.

Dr. Shakespeare admitted, as absolutely established, the power of the tubercle bacillus, under favorable conditions, to produce a genuine and virulent form of tuberculosis. He did not admit that it had been positively demonstrated that no other agent might also be capable of producing the disease; on the other hand, he denied that it had been satisfactorily proved that any other agent was capable of exciting tuberculosis. He believed the proof strong that, under certain favorable conditions, tuberculosis was an infectious disease, and that, at least frequently, the infecting agent was the tubercle bacillus. He saw no valid reason to deny that, under certain favorable conditions, tuberculosis might be conveyed from person to person, and in this sense he termed a contagious disease. Whether or not the tubercle bacillus were regarded as the only agent capable of exciting tuberculosis, its virulence was certainly incomparably greater than that of any other known agent. He therefore failed to appreciate the wisdom or the logic of those who, admitting the virulent qualities and propagative power of the tubercle bacillus, yet, because of a lingering suspicion or even of a decided belief that other agents could produce this terrible disease, would still decline to guard against possible infection or contagion. He regarded the tubercle bacillus, when present, as an infallible sign of the presence and activity of the tuberculous process. On the other hand, its absence, unless after repeated and long-continued searches by competent observers, did not positively warrant a negative conclusion. He therefore saw in the tubercle bacillus an important means of differential diagnosis in obscure cases. From its reported presence in some cases earlier than the physical signs could possibly determine a diagnosis of phthisis, he was inclined to think that it might become of inestimable value to the skillful practitioner to forewarn him of the beginning of that formidable malady which, if curable at all, must be combated from the very onset.

Dr. Woodbury said that at least two distinct questions had been submitted for discussion: Was consumption contagious, and Was the Bacillus tuberculosis the efficient and only cause of consumption? One of these was not necessarily the complement of the other. Consumption might be contagious without being caused by a bacillus, and bacilli might cause consumption without rendering it contagious. The first question he thought should be decided by clinical experience, the second by clinical experience with the aid of morbid anatomy and mycology. Time would permit only a very brief presentation of the arguments in favor of the views which he held, and he therefore would at once state his conviction, and he believed the experience of others would agree with his own, that pulmonary consumption as ordinarily met with was not a contagious disease. Since the definition of a disorder must be made from the clinical picture presented by the majority of cases, he would say that the typical case of consumption did not present any evidence of possessing a contagious character. The question as to the communicability of consumption under exceptional circumstances he regarded as a very different one from the former. Meningitis or nephritis might be communicated under peculiar conditions, but this would not warrant the clinical teacher in describing them as contagious, at least in any ordinary acceptation of the word. He had seen a number of cases of consumption which had occurred in members of one family living under the same conditions, but had never met with a single case where the evidence of contagion was conclusive. Even cases of apparent communication from husband to wife, or *vice versa*, could be satisfactorily explained to his mind on other grounds than those of direct transfer of the disease by organic or organized particles. The susceptibility to phthisis might be native or acquired; it could not be transmitted by particulate infection.

With regard to the ætiology of consumption, it would appear that there were several varieties of the disease which were indistinguishable by ordinary physical signs. In the first place, there were two classes of cases which stood apparently identical, differing in the microscopical characters of the sputum; one contained the alleged *Bacillus tuberculosis*, the other did not. This led us to a classification of bacillary and non-bacillary tuberculosis. In the latter class of cases, in addition to syphilitic phthisis, pulmonary actinomycosis, and zoöglœic tuberculosis (a form of mycosis recently described by Malassez Vignol), there were included cases of ordinary pulmonary phthisis, but *minus* the bacillus. In the first class, therefore, the question arose, "were the bacilli necessarily the cause of the morbid phenomena?" He thought that they were not essential, (1) because it had been shown that consumption might be due to other causes and pursue its course without their appearance, and (2) because they were apparently not a necessary element of tubercle. The bacilli had undoubtedly a certain diagnostic and prognostic value, but their appearance could be accounted for on the hypothesis of their being a mere concomitant of pulmonary consumption, even though it could be shown that they increased its fatality. He was surprised that, with such abundant opportunities for observation, clinical teachers had not been able to convince the world or themselves that consumption was contagious until they were shown something under a microscope. He was more than surprised that Professor Austin Flint had announced his adherence to the new doctrine, that pulmonary consumption was due to the *Bacillus tuberculosis*, and arose in no other way.

Dr. GEORGE HAMILTON said that, after a practice of more than half a century, he had seen no case of pulmonary consumption that could rationally be attributed to contagion. In two or three families, where several members were affected with this disease, attempts had been made to refer it to contagion, but without any sufficient proof. It was to be borne in mind that great repugnance sometimes existed in a family to admitting an hereditary tendency to this affection, scrofula, and certain other maladies.

Dr. DUXMING said that on the question, "Whether or not simple inflammation of serous membranes could lead to tuberculosis in the non-scrofulous," he would say that he had the

* "Journ. of the Am. Med. Assoc.," February 16th, from "Archives de physiologic."

notes of a case in which the post-mortem proved death to have been caused by phthisis pulmonalis, in which the primary trouble seemed to be the fracture of two ribs on the right side. While both lungs were involved, the pleuritic adhesion of the right side was almost entire. An intimate acquaintance with the family, both before and since the death of this patient, had failed to show any sign of tubercular trouble, and, so far as he knew, none of this connection had died of the disease.

PHILADELPHIA CLINICAL SOCIETY.

Meeting of May 23, 1884.

Dr. HENRY BEATES, JR., in the chair.

A PUERPERAL CASE WITH NUMEROUS COMPLICATIONS was reported by Dr. MARY WILLETS. Mrs. H., aged thirty, a primipara, after a normal delivery did well for twelve days. Then, after pain in the back and limbs and chilly sensations, she had a rise of temperature and was attacked with nausea and vomiting. The temperature continued high for two weeks. There was nothing to account for the elevated temperature except a laceration of the cervix uteri and some tenderness around this point. On the twenty-fifth day after delivery the patient complained of pain in her left leg. For more than a week there were pain and swelling both above and below the knee; the pain was greatly increased on pressure and on attempts at extension of the limb, and was in the course of the femoral vein, but careful examination failed to reveal anything abnormal. On the forty-third day, the patient having recovered sufficiently to go down stairs, there was a sudden attack of well-developed mania, the patient being violent at first, but subsequently merely loquacious. This continued two days, and then gave place to somnolence, which lasted five days, after which convalescence began.

Dr. E. E. MONTGOMERY remarked that the case was unusual from the lateness at which the fever appeared. During his present term at the Philadelphia Hospital measures had been instituted to prevent the contact of septic matters with the parturient parts. A solution of corrosive sublimate, one to two thousand parts, was used to sponge the parts with after the expulsion of the placenta, and cloths saturated with the solution were kept in place by absorbent cotton, oiled silk, and a tailed bandage. There had been but three cases of septicæmia during the present quarter.

Dr. ALBERT H. SMITH said the local examination which the reader of the paper had very properly made solved the whole problem. Together with the symptoms, it showed the case to be one of pyæmic—not septicæmic—poisoning in a woman feeble and unable to resist the absorption of pus. Septicæmia could not arise after all open surfaces had become purulent. The examination was valuable, and he thought it would be better if more care was generally exercised in this direction.

CLINICAL PHENOMENA FOLLOWING THE PUERPERAL STATE IN TWO CASES.—Dr. PHILIP M. SCHIEDT read a paper with this title. In the first case he was suddenly summoned to the bedside of a woman, aged about thirty-five, whom he found comatose, with the pupils undilated, the head drawn to the left side and flexed on the chest, a clammy sweat on the forehead, and the pulse imperceptible. The heart sounds were faintly detected, and the body and extremities were warm. Restoratives and artificial respiration were without avail, and the patient soon expired. The history of the case, subsequently ascertained from the family physician, was as follows: She had been delivered of a child after a natural labor two weeks previously, which was followed by a normal convalescence unaccompanied by fever or offensiveness of the lochial discharge. Her babe was healthy, and was